Kingdom Builders Publications LLC

ISBN: 979-8-218-73085-7 Soft Cover

Printed in USA

Authored by
Huizhen Mount

Kingdom Builders Publications LLC

Quantum Simultaneity

Huizhen Mount

Table of contents

Introduction

Digital bridge between I Ching symbols and Western science 3

To be inspired and to inspire 18

Points of controversies 21

Self reflection 25

Introduction

Digital Bridge Between I Ching Symbols And Western Science

This small book is going to be focused on Entirety in the sense of Consistency. Based on my previous two publications titled "The Universal Law – Creation and Evolution in Cosmic Double Helix " and "Living in the Cosmic Computational Genome ", I noticed myself on some issues for improvement, that is Simplicity needs Consistency which lacked somewhere in some chapters. I am intending to bring clearer illustrations with common feature: digital significance in macro and micro DNA. Such digital significance is what I call the Bridge between I Ching symbols and Western science.

Without this discovery, my research on the nature of Cosmic DNA and individual DNA could face a big hurdle. In a much precise expression, this bridge took me to the discovery on deeper meaning in Einstein's special theory of relativity : $E = (M-M)C^2$ can be translated into the following expression: The concept of energy can be expressed as " Yang ___ ", concept of mass can be expressed as " Yin _ _ ". Their digital expression is in common form : 1 for energy, Yang ; 0 for mass, Yin. Here is an illustration for those application in digital term.

First, translate such digital form into Semiconservative replication in 1s and 0s:

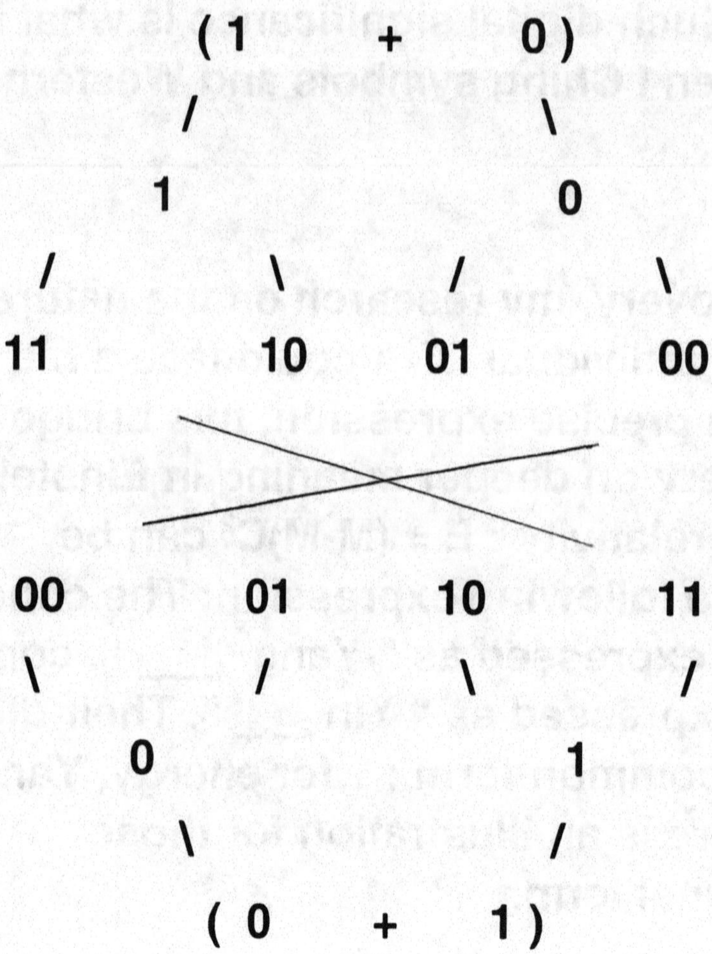

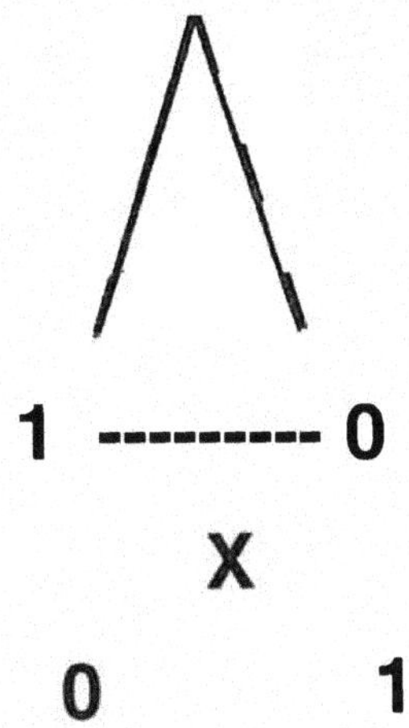
1 --------- 0
X
0 1

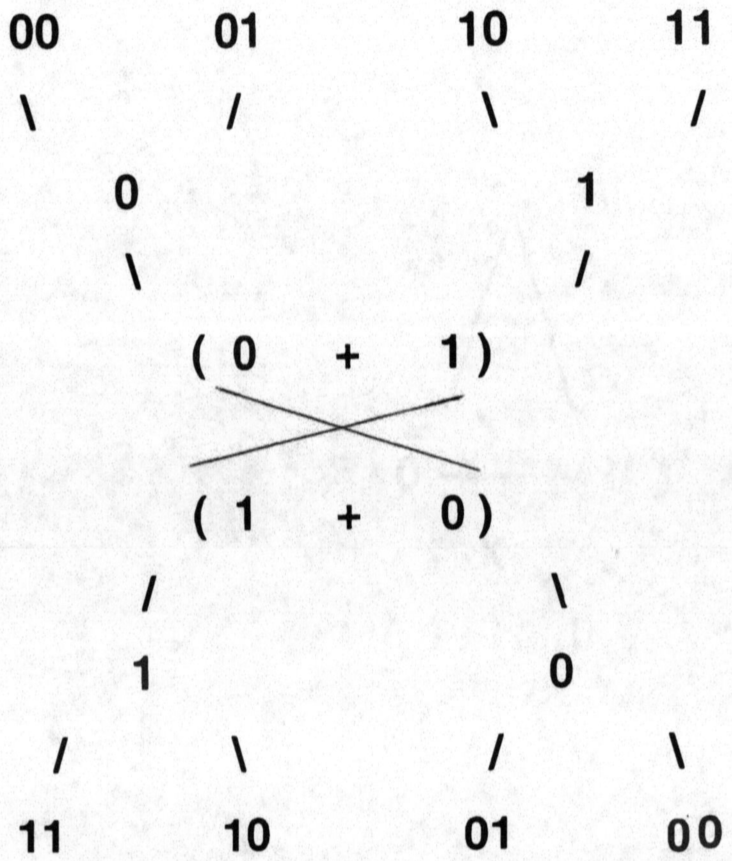
00 01 10 11
0 1
(0 + 1)
(1 + 0)
1 0
11 10 01 00

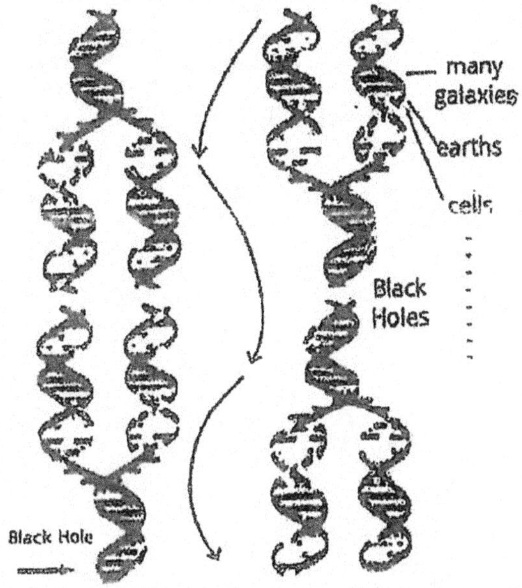

Migration of our DNA through the path of double helix

E= (M+M-) C²

M+ is matter, M- is antimatter

Evolution and Creation side by side in general theory on the formati of genetic codes

Tó express identical idea, this illustration can be applied in the term of a cell, the universe. Draw an image of cell's replication.

The entire structure of the transformation between energy and mass is the transformation of Yin and Yang. Here is the translation:

```
          ( E     +     M )
         /                 \
        E                   M
      /   \               /   \
    EE     EM           ME     MM
```

(energy and mass in semi conservative replication)

The following pages display the reversed expression:

Reverse engineering >Time reversal>Negative time. The ultimate significance is in Quantum curvature that bring curved space time in the sense of Quantum Simultaneity.

By crossing the bridge, digital science unfolds itself with much more profound application into next step:

The Unified Theory in Cosmic Information.

Image 1

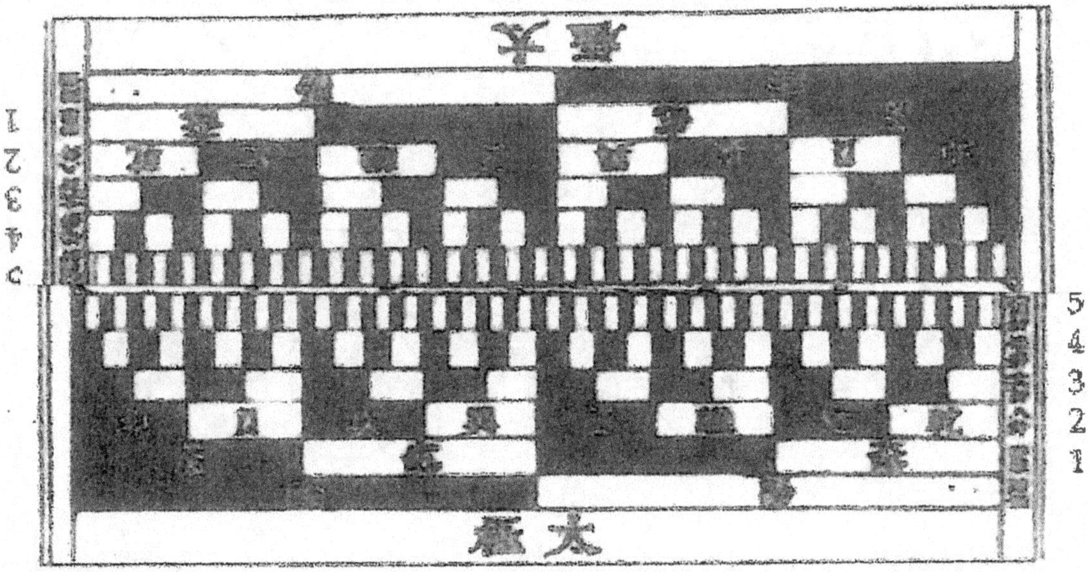

Image 2

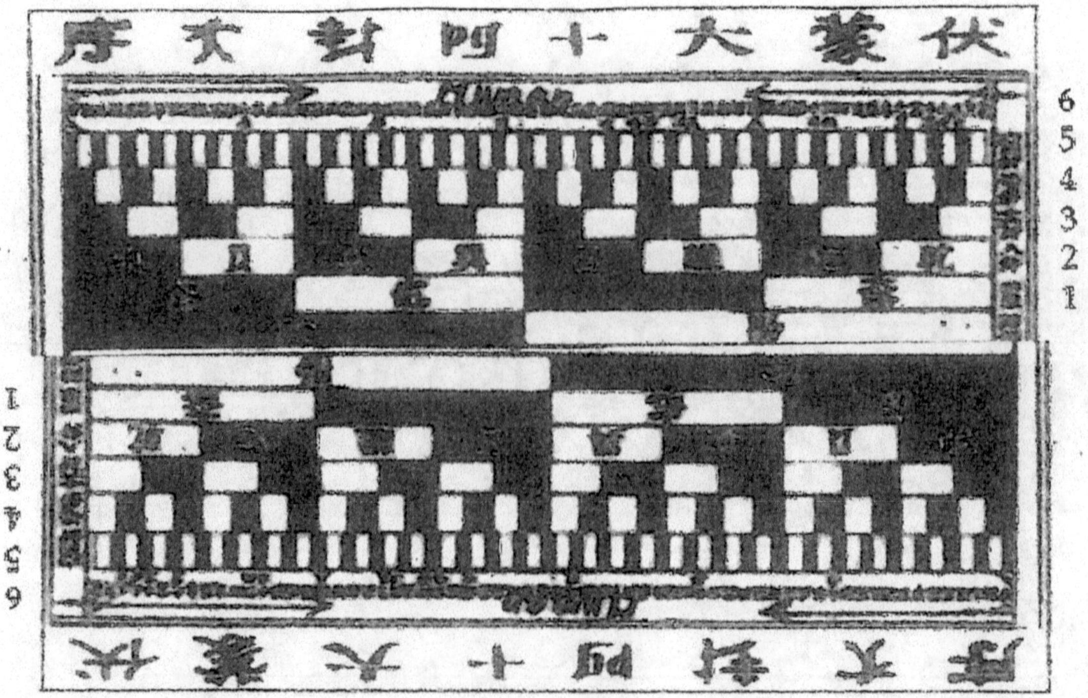

Image 3

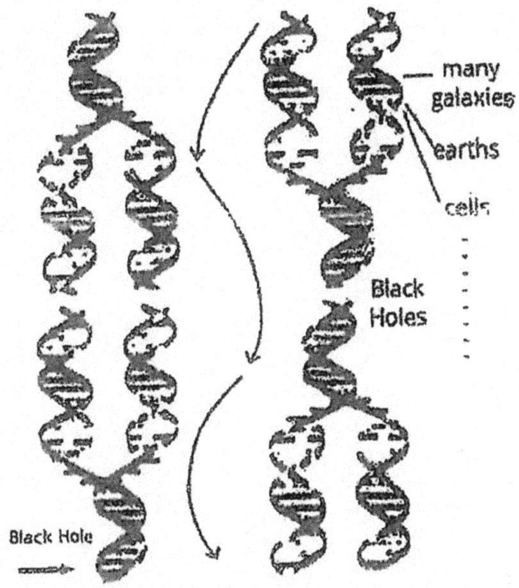

Migration of our DNA through the path of double helix

E= (M+M-) C2

M+ is matter, M- is antimatter

Evolution and Creation side by side in general theory on the formati
of genetic codes

Tó express identical idea, this illustration can be applied in the term of a cell, the universe. Draw an image of cell's replication.

Image 4

I noticed myself on some issues for improvement, that is Simplicity needs Consistency which lacked somewhere in some chapters.

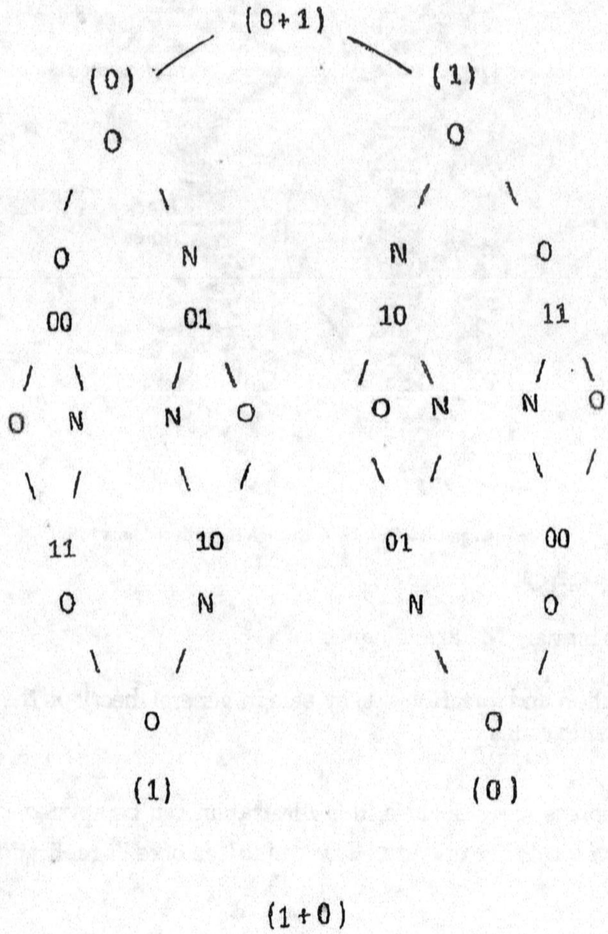

In this reversed geometric expression, circle is to express the curved space time. The line between two triangles has dual meaning, that is

---

X

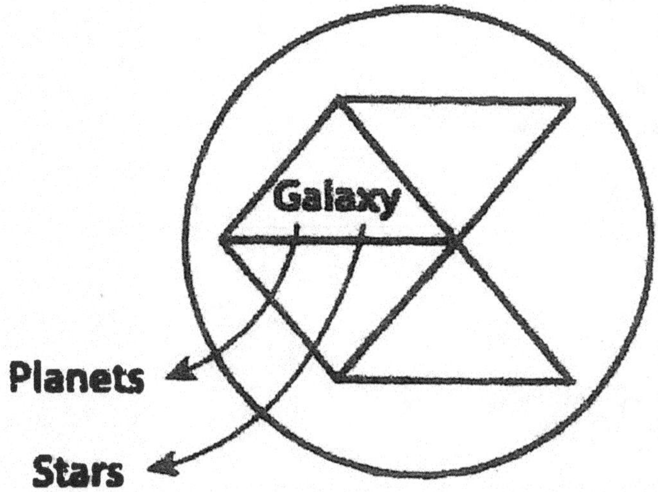

Image 5

Mass and energy are in consolidation for contraction. And it's digital expression is : 1 for energy , 0 for mass. Consolidation of 1 and 0 is such singularity

```
O  N   N   O  O  N  N  O          M =E
    O   N     N    O              ____
    (0) O        O(1)              C2
         \       /
          (0 + 1)
            \/
------------/\----------------------------- boundary
          ( 1+0 )                 E = MC2
         /       \
    (1 ) O       O (0)
    /    \     /    \
  O       N   N      O
 /   \ /  \/  \  /   \
O  N  N  O O  N  N   O
```

Expansion is the reverse of contraction

To Be Inspired And To Inspire

Throughout my research process, Einstein's equation $E=MC^2$ inspired me on various subjects, such as Cosmic double helix, hidden digital nature by linguistic comparison in the term of I Ching symbols with Einstein's scientific language. However, due to lack of research information on the equation $M=E/C^2$, I sensed invisible consequences caused to our lack of knowledge on the fuller scale on the astronomy and physics. I had attempted to bring these two equations into one in the term of quantum simultaneity:

Can modern vocabulary like " encrypt " and "decrypt" be applied to computing language for " gene chip " ?

Points of Controversies

Following topics listed were controversial at the beginning of my research, as inspiration came to me from controversial topics that are very challenging for me to tackle academically and practically. My approach to those challenges is Not to avoid discussion, but bring them to a long road of research, regardless how little or how much I can achieve.

1. **What does $E=MC^2$ have to do with Cosmic double helix?**
2. **Modern science has reached a very advanced stage, how comes $M=E/C^2$ has not been broadly researched?**
3. **Can unification theory be found in Einstein's general and special theory of relativity with a form of quantum simultaneity?**

   How are galaxies connected to one another?
4. **Life beyond our own galaxy.**
5. **Is digital structure able to bring a clear and simplified information for Cosmic worlds?**

6. **How is individual DNA of any species or a person is related to Cosmic DNA?**
7. **Under cosmic DNA , how do cells communicate with each other?**
8. **How do interplanetary structure bring invisible galactic life into a possible digital view?**
9. **There are so many kind of computer chips with semiconductor tech for smart phones, tablets, auto machines, wireless, why can't there be computer chip for genetic science?**
10. **Can unification theory be explored in digital form?**

11. **Is the equation $M=E/C^2$ a valuable topic for science research?**

12. **The foundation of unification theory is in self digitized Cosmic DNA where macro and micro genome reside. How can any micro genome be located inside the space time of macro genome?**

13. **"A looking glass" is the analogy I apply toward the expanded view for Cosmic DNA from our living cells.**

14. Can Cosmic computational genome be considered as Cosmic memory ?

Self Reflection

Quantum simultaneity paves a clear path for unification theory that I have been searching. Einstein's general and special theory of relativity taught me profound philosophical significance along with physics and astronomy and so on. And this cannot be achieved without comparative linguistic analysis from I Ching symbols.

Thus, the concept of unification theory brings the Universal language for understanding the nature of information science at macro and micro scale.

www.ingramcontent.com/pod-product-compliance
Lightning Source LLC
LaVergne TN
LVHW080528210926
848481LV00022B/3653
9798218730857